MOON

A Guided Journal

LEIGH PATTERSON

Clarkson Potter/Publishers
New York

LISTS

An ordinary life was an obscure life, if we can extend the meaning of obscure to mean covered up by dailiness, glorious dailiness, shameful dailiness, dailiness that is difficult to figure out, that is not always clear until a long time afterward. Obscure: not readily noticed, easily understood, or clearly expressed. Which is a pretty good definition of life.

—Mary Ruefle, "I Remember, I Remember"

I REALIZED RECENTLY that the narrative I was telling others in casual conversation needed adjusting. "Hi, how have you been? What's new?" a friend would ask—someone I care about but haven't seen in a while. "Mostly work," I'd respond.

Is that really it? Just work?

An egg and toast, sitting at my desk, headphones in my ears, guilt about forgetting to call my mom back, daydreams, a lot of abstract wants. How do I refine, how do I redefine?

THE MOON LISTS

The Moon Lists are a set of questions to inspire reflection on the recent past. These questions, prompts, and lists are an exercise in drawing the connection between memory and experience, a reminder that daily choices matter, habits are defining, and objects can hold layers of significance.

The cycles of the moon have historically been regarded as a sort of compass for self-reflection; with the unending ebb from night to light, the moon acts as a routine reminder that we are guided by months and seasons, a system of order and natural rhythm that exist outside of ourselves. Traditionally and across cultures the phases of the moon are regarded with different types of meaning, from new beginnings to intention setting to times of surrendering acceptance.

As the moon cycles through its phases, you will be asked to check with yourself, once a week, to reflect on your recent experiences.

Every week in this journal provides a list-making prompt or memory-gathering touchstone for self-reflection. There are different ways to consider the present, from the objects and themes that are front-of-brain to the sensory details that have filled in the spaces in between moments. There is also space to catalog a more analog scrapbook of relevant ephemera, paper, or to recall notes or conversations that played a part in the last week. Take stock of what you recently acquired; this is a space for a souvenir from the week. A receipt, a leaf, a parking ticket, a shred of a pistachio shell. Tape it into your book.

At the end of each month is a set of questions constructed to look back on the last four weeks as a collective whole. You can define "the last four weeks" however you want: maybe you complete your list on the first of each month or maybe you just jump in whenever you can. Maybe you reflect on the full or new moon. Do what works for you.

Maybe you share and talk through your lists with a friend or partner. Maybe you develop your own ritual surrounding self-reflection (lighting a candle or a stick of special incense).

Maybe you do it outside or at dinner, or at the park. Maybe it is spontaneous. Maybe it's written down; maybe spoken aloud.

I do think it's helpful to flip ahead to the prompts as you start your month to have them in the back of your mind as you go about your weeks . . . or if the prompts of a sequential month don't feel right, skip ahead. The goal here is just to set aside time for weekly and monthly reflection, however that works best for you; it's a reminder to notice something you might not otherwise and readjust your framing.

Choosing the way you observe, reflect on, and document what's going on at present is a reminder: you are in control of your narrative. Call out the small moments that add meaning to your life, that fill you up on a deeper level, and seek out more of those experiences. This is the truest, most timeless form of self-care. It costs nothing other than your attention, asking that you live with your eyes open.

Artists, poets, whatever you want to call those people whose job is "making" take in the commonplace and are forever recognizing it as worthwhile.

I think I am always collecting in a way, walking down a street with my eyes open, looking through a magazine, viewing a movie, visiting a museum or grocery store. Some of the things I collect are tangible and mount into piles of many layers and when the time comes to use those saved images I dig like an archaeologist and sometimes find what I want and sometimes don't.

—Sister Corita Kent

THE MOON LISTS

ON CIRCLES

Circles are symbols of wholeness, no start and no finish, the unknowable origin of time. The eternal. Source of life, destination of uncertainty. The circle signals repetition, movement within constancy, gateways, the act of centering. It is a signifier of equality, equidistance. Oneness.

In the same way that meditation practices incorporate mantras for bringing you back to your personal center, think of these motifs as symbols hidden in plain sight, reminders to seek spaciousness, step outside of your experience, and take note. The moon can be your mantra.

CONSIDER

Evaluate "means" by assigning value to "ends."

0 / 54

THE MOON LISTS

CONSIDER

Each week on the left-hand side of your journal, you'll see what we call "Consider Cards." Each card is an invitation to do just that—consider finding your own meaning. Maybe one week it pertains to your present state of mind. Or maybe it will prompt you to ask your own questions, wonder about possibility, or observe the multidimensionality of every moment.

Make a recognition in silence.

1 / 54

WEEK ONE

TAKE A RECENT INVENTORY

1. A sight:
2. A smell:
3. A sound:
4. A touch:
5. A taste:

Unknown circular chart,
ca. 1500.
The New York Public Library Digital Collections.

1.

2.

3.

4.

5.

Maybe it's meant to be a difficult month.

2 / 54

WEEK TWO

WHO IS THE YOU OF NOW?

A survey of present interests, curiosities, recurring themes . . . a space for a short list of some front-burner items:

1. Puglia, Italy
2. Orange blossom in the air
3. Sparkling water
4. Kenny Rogers and Dolly Parton's "Islands in the Stream" duet
5. *Mono no aware*

1.

2.

3.

4.

5.

Egyptian Papyrus Lids,
ca. 1336–1327 B.C.
Metropolitan Museum of Art.

Take up a lot of space.

3 / 54

WEEK THREE

A SPACE FOR SCRAPS

A space to preserve. Tape or insert something (or things) you've acquired this month, a tactile reminder of the present.

The Past Month

1. STORY
   What was the best story of the last month? Maybe it was something that happened to you, maybe it was something that was told to you or that you observed.

2. CONNECTION
   With what (or whom) did you connect?

3. LOSS
   What was lost, what are you mourning? Maybe it is minor.

4. ABUNDANCE
   What was plentiful?

5. OTHER WORLDS
   What was a moment where you glimpsed another world: a life or a reality you are not part of? Maybe it was the result of eavesdropping, a toe dipped into an unfamiliar scene, or just acknowledging a path you didn't choose.

6. NATURE
   An encounter with and/or in the outdoors. Connection with the natural world.

7. MINOR SECRETS
   Describe something you did in private. Perhaps it's not really a secret, but something that never occurred to you to share . . .

8. CULTURE LIST
   What was read, watched, seen, listened to? And consider the ratio between the mediums.

1.

2.

3.

4.

WEEK FOUR

5.

6.

7.

8.

MINOR SECRETS

THE BATTERIES IN the smoke detector went out and somehow I could not figure out how to replace them. Normally, I am decent at following instructions but have admitted utter defeat in the hands of a tiny, chirping, relentless smoke detector. I took it off the wall and hid it in a cabinet. Will live a little riskier from here on out.

CONSIDER

Sometimes the magic works and sometimes it doesn't.

4/64

WEEK ONE

TAKE A RECENT INVENTORY

1. A sight:
2. A smell:
3. A sound:
4. A touch:
5. A taste:

An inventory of tools from the "20th Century Catalogue of Supplies for Watchmakers, Jewelers, and Kindred Trades," 1899. *The Library of Congress.*

1.

2.

3.

4.

5.

Do a task
in secret.

5 / 54

WEEK TWO

WHO IS THE YOU OF NOW?

A survey of present interests, curiosities, recurring themes . . . a space for a short list of some front-burner items:

1. Beeswax candles
2. A weighted glass tumbler—used for a specific drink
3. Athens, Greece
4. Oolong tea
5. Mysteries

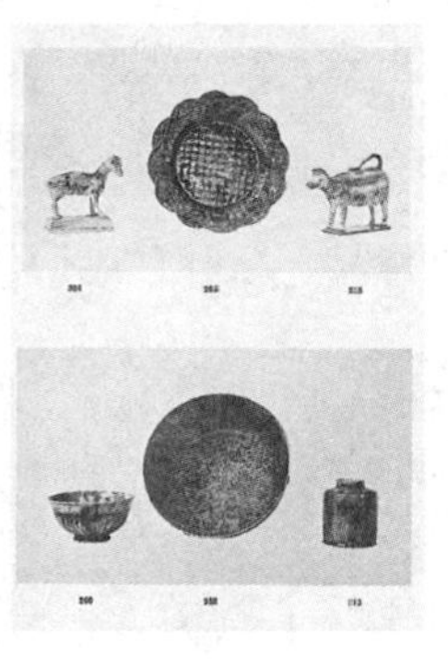

1.

2.

3.

4.

5.

Tortoise-Shell Circular Tea Caddies, Bowls, Jug, And Animal Sculptures, 1915. *The New York Public Library Digital Collections.*

Identify a medium that moves you.

6 / 54

WEEK THREE

A SPACE FOR SCRAPS

A space to preserve. Tape or insert something (or things) you've acquired this month, a tactile reminder of the present.

The Past Month

1. OBJECT
   An object (New? Old?) that played a role in your life this month.

2. SIMPLE NOURISHMENT
   What simply satisfied you? Maybe it was a ten-minute phone call with a friend. Maybe it was a perfectly timed bath. Maybe it was a sandwich.

3. FUNNY
   What was a funny thing that happened? When did you laugh most deeply? When did you smile like a fool? Recount the experience.

4. MYSTERY
   What happened that you can't explain?

5. ALONE TIME
   What role did alone time play? Recount an experience.

6. PEOPLE TIME
   What role did others play? When and how did you connect in the way you needed most?

7. SLEEP
   Any dreams?

8. CULTURE LIST
   What was read, watched, seen, listened to? And consider the ratio between the mediums.

1.

2.

3.

4.

WEEK FOUR

5.

6.

7.

8.

SIMPLE NOURISHMENT

AT THE BEGINNING of the month I traveled to the southern coast of Italy. The first night I arrived a few hours later than I'd planned, a bit delirious from jetlag and airport delays. The kitchen had closed at the tiny hotel where I was staying and the nearest restaurant was nearly an hour away. As I settled into my room there was a knock on my door; the owner of the property was standing outside, holding a plate of homemade pasta, simply dressed with olive oil and pepper. A small act of kindness and actual nourishment that was one of the most memorable meals and moments of the entire trip.

Do not be a magician—be magic!

—Leonard Cohen

7 / 54

WEEK ONE

TAKE A RECENT INVENTORY

1. A sight:
2. A smell:
3. A sound:
4. A touch:
5. A taste:

Gold Iranian coin, ca. 1099.
Metropolitan Museum of Art

1.

2.

3.

4.

5.

Some days
you feel flat;
some days you
feel fine.

WEEK TWO

WHO IS THE YOU OF NOW?

A survey of present interests, curiosities, recurring themes . . . a space for a short list of some front-burner items:

1. Hydrangeas growing in the wild
2. A coconut parfait
3. Inverness, California
4. Quilted textures
5. A bare face

1.

2.

3.

4.

5.

Goldstone Antenna
Construction Model, 1963.
NASA/JPL-Caltech

CONSIDER

Recall a lost weekend.

9 / 54

WEEK THREE

A SPACE FOR SCRAPS

A space to preserve. Tape or insert something (or things) you've acquired this month, a tactile reminder of the present.

The Past Month

1. CHANGE
   A change you enacted with intention, or one that came from the outside.

2. ART EXPERIENCE
   An encounter with art (in any form).

3. GIFT
   Something you gave or received.

4. NOSTALGIA
   What was felt more deeply because it took you back to your past?

5. RESIST
   What did you resist?

6. SURRENDER
   What did you surrender to?

7. PHYSICAL SELF
   Take a scan. What happened to your body? How do you *feel?*

8. CULTURE LIST
   What was read, watched, seen, listened to? And consider the ratio between the mediums.

1.

2.

3.

4.

WEEK FOUR

5.

6.

7.

8.

SURRENDER

As a wedding gift I received a set of very nice French paring knives. I'd wanted them a long time. Wooden handles and a carbon steel blade. Lightweight but sturdy. To my dismay, the first time I used them they immediately started to discolor, literally oxidizing before my eyes. Apparently, they can't get wet. Parenting a knife isn't something I anticipated; I've accepted that sometimes I ruin nice things.

CONSIDER
The answer you're looking for is probably in a book somewhere.
10 / 54

WEEK ONE

TAKE A RECENT INVENTORY

1. A sight:
2. A smell:
3. A sound:
4. A touch:
5. A taste:

Subgeometric Plate,
680–660 B.C., Terracotta.
The J. Paul Getty Museum.

1.

2.

3.

4.

5.

It doesn't have to be serious!

11 / 54

WEEK TWO

WHO IS THE YOU OF NOW?

A survey of present interests, curiosities, recurring themes . . . a space for a short list of some front-burner items:

1. Crewneck sweatshirts
2. A shot of ginger
3. John Berger
4. Going analog
5. Spanish lessons

1.

2.

3.

4.

5.

Incantation bowl,
Mesopotamia,
ca. 5th–9th century.
The New York Public Library Digital Collections.

How can you reel in?

12 / 54

WEEK THREE

A SPACE FOR SCRAPS

A space to preserve. Tape or insert something (or things) you've acquired this month, a tactile reminder of the present.

The Past Month

1. JOY
   A time of celebration, deep happiness, or that made you feel emotionally full.
2. ABSENCE
   What was missing last month?
3. LIMIT
   When were you (or were you ever) pushed to your limit last month?
4. THEMES
   Were there any themes last month?
5. CLOTHING
   What did you wear? Are there particular garments that are having a moment for you?
6. PARE BACK
   Did you simplify something?
7. PROPORTION
   A specific moment that reminded you of the scale of the universe, that you are part of a greater whole. An outdoor shower in the mountains. A particular drive at night. Reading a specific quotation at the right time.
8. CULTURE LIST
   What was read, watched, seen, listened to? And consider the ratio between the mediums.

1.

2.

3.

4.

WEEK FOUR

5.

6.

7.

8.

THEME

THE WORD OF the month is *transition.* I can sense a change in my own perspective, an awareness of some qualities in my surroundings and in my behaviors that are holding me back. I feel a little dulled, tired. I want to be sharp. I want to be awake. I opened my desk drawer yesterday and saw a bright pink swatch of tulle that I've been keeping because the color is so shocking. I pinned it on my wall as a reminder: Eyes open.

CONSIDER

Tolerate the ambiguous.

13 / 54

WEEK ONE

TAKE A RECENT INVENTORY

1. A sight:
2. A smell:
3. A sound:
4. A touch:
5. A taste:

Instruction in the Elements of the Art of Astrology, 1934.
The New York Public Library Digital Collections.

1.

2.

3.

4.

5.

What do you wish you had the manual for?

14 / 54

WEEK TWO

WHO IS THE YOU OF NOW?

A survey of present interests, curiosities, recurring themes . . . a space for a short list of some front-burner items:

1. Raisons d'être
2. Joni Mitchell in Ibiza
3. Old friends
4. Citron
5. Campari

1.

2.

3.

4.

5.

Looking glass in silver and bronze, designed and executed by Elkington & Co., 1880–1883.
The New York Public Library Digital Collections.

Eat something soggy.

15 / 54

WEEK THREE

A SPACE FOR SCRAPS

A space to preserve. Tape or insert something (or things) you've acquired this month, a tactile reminder of the present.

The Past Month

1. REARRANGE
   Did you put something in a new order? Maybe something suits you better at a different time of day; maybe you organized a desk drawer.

2. PRIDE
   When were you proud? Of whom, and for what?

3. UNEXPECTED PAIRINGS
   What nonobvious combinations had a role: two random songs, played sequentially and on repeat; a peach and a tomato, together in a salad.

4. TIME OF DAY
   What times felt significant: Dawn or dusk? A regular hour that brings lull or excitement? A time you looked forward to or anticipated?

5. VEG
   What circumstances prompted you to be lazy? Was it nourishing or numbing?

6. CHALLENGE
   What was a question you found yourself asking?

7. FAILURE
   What didn't go as planned?

8. CULTURE LIST
   What was read, watched, seen, listened to? And consider the ratio between the mediums.

1.

2.

3.

4.

WEEK FOUR

5.

6.

7.

8.

UNEXPECTED PAIRINGS

I RECEIVED A beautiful handmade ceramic tea strainer as a gift. It's beautiful ivory, hand-formed, and just irregularly shaped enough to feel interesting. The day I received it, I randomly stuck a garlic bulb in it; it fit perfectly. I've yet to actually use the strainer to make tea because I'm too pleased with how it looks as an oddly pleasing art object.

Can you take an idea and release it back into the world?

WEEK ONE

TAKE A RECENT INVENTORY

1. A sight:
2. A smell:
3. A sound:
4. A touch:
5. A taste:

Moon Crater, late 1850s.
The J. Paul Getty Museum.

1.

2.

3.

4.

5.

Ask for help.

17 / 54

WEEK TWO

WHO IS THE YOU OF NOW?

A survey of present interests, curiosities, recurring themes . . . a space for a short list of some front-burner items:

1. Blank slates
2. The writing of Tara Brach
3. Surrealists
4. Taos, New Mexico
5. Swimming

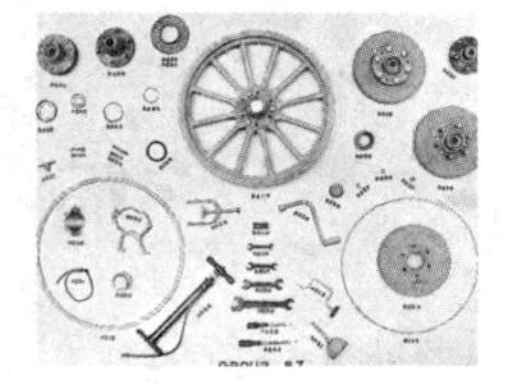

1.

2.

3.

4.

5.

Rambler group 87, wheel and tool equipment, 1903–1916. *The New York Public Library Digital Collections.*

Maintaining wonder is rigorous work.

18 / 54

WEEK THREE

A SPACE FOR SCRAPS

A space to preserve. Tape or insert something (or things) you've acquired this month, a tactile reminder of the present.

The Past Month

1. SURPRISE
   What was shocking (perhaps delightfully so)?

2. VULNERABILITY
   When did you feel vulnerable to envy? Feeling competitive?

3. CHANGE
   A change you enacted with intention, or one that came from the outside.

4. RITUAL
   What did you repeat with intention or delight? What practices serve you from their repetition?

5. INDULGENCE
   When were you indulgent? When were you self-indulgent?

6. EYE OF THE BEHOLDER
   What was a moment of unexpected beauty, perhaps seen (or understood) only by you?

7. BRAVERY
   What was a moment that took courage?

8. CULTURE LIST
   What was read, watched, seen, listened to? And consider the ratio between the mediums.

1.

2.

3.

4.

WEEK FOUR

5.

6.

7.

8.

RITUAL

I WEAR AN oil fragrance but the container it comes in is kind of ugly, so I started religiously decanting the contents into a miniature jade vial. Dabbing it on my wrists, spinning the heavy stone cap back onto the bottle, holding the cold jade in my hand . . . currently seeking more sensations that inspire opulence.

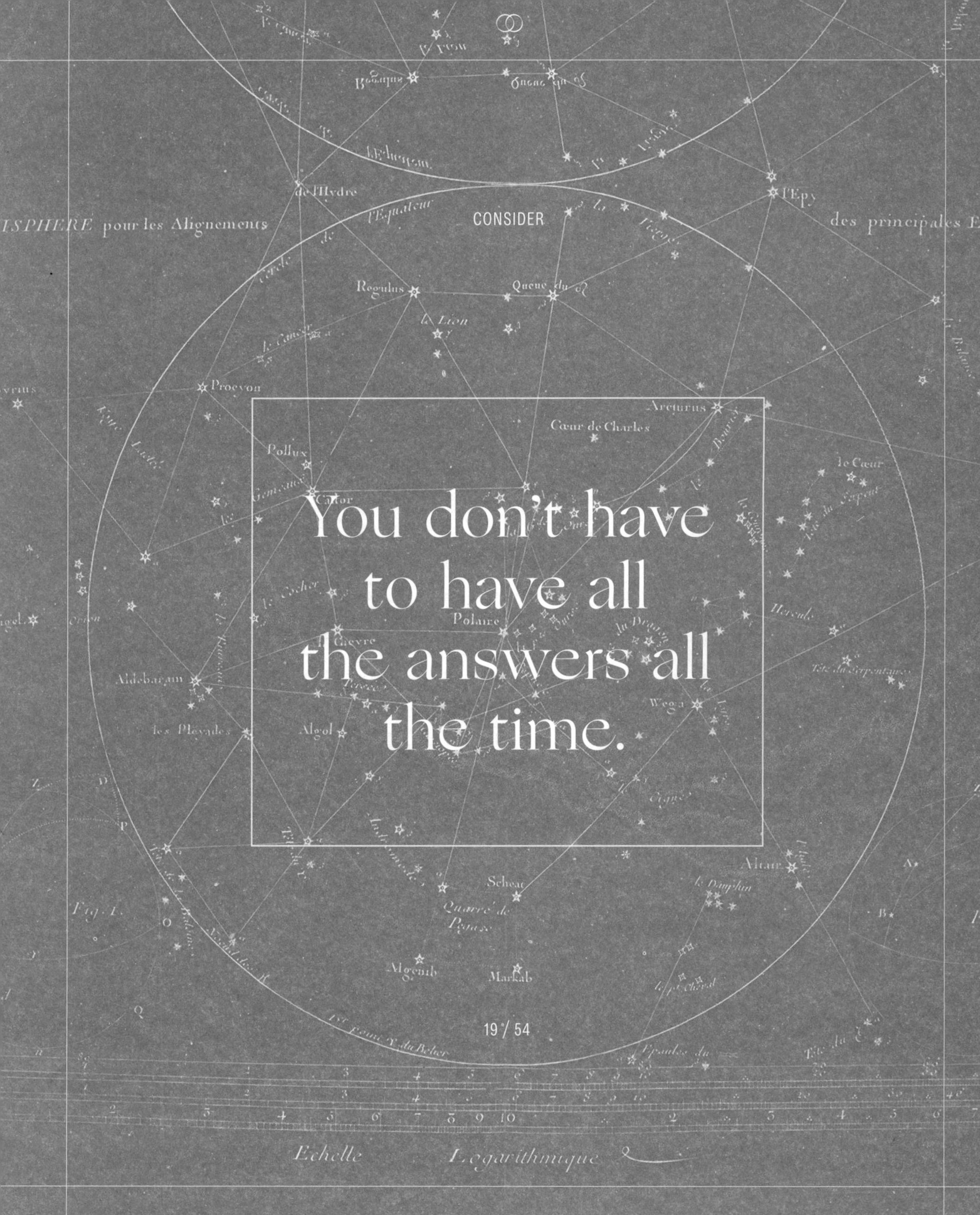

You don't have to have all the answers all the time.

19 / 54

WEEK ONE

TAKE A RECENT INVENTORY

1. A sight:
2. A smell:
3. A sound:
4. A touch:
5. A taste:

"Planisphere & figures," 1795.
The New York Public Library Digital Collections.

1.

2.

3.

4.

5.

Protect your time.

WEEK TWO

WHO IS THE YOU OF NOW?

A survey of present interests, curiosities, recurring themes . . . a space for a short list of some front-burner items:

1. Island time
2. Form over function
3. Integrity
4. A pinky signet ring
5. Heroes

1.

2.

3.

4.

5.

Practice for the Florida State Tarpon Club, Wakulla Springs, 1949. *The State Library and Archives of Florida.*

Untangle the issue. Rearrange something.

21 / 54

WEEK THREE

A SPACE FOR SCRAPS

A space to preserve. Tape or insert something (or things) you've acquired this month, a tactile reminder of the present.

The Past Month

1. NARRATIVE
   You run into a friend you haven't seen in a while—"What's been going on with you?" What's the story you're currently telling about yourself, to yourself (and to others)?

2. RETURN
   What did you revisit last month? A song you played over and over. A film you rewatched. A breakfast rut.

3. CONCLUSION
   What is something that came to an end this month? Are you happy to see it go?

4. EMOTE
   Crying, yelling, acting out. What emotions took focus (or perhaps took control)?

5. PROCRASTINATE
   What did you put off? Did you procrastinate small tasks or broad categories?

6. WONDER
   What moments crossed the threshold in making you wonder? When did you feel amazement at the world around you or experience a marked break from your routine?

7. SENSORY TABLEAU
   What were some sense-driven experiences: textures, colors, smells, tastes?

8. CULTURE LIST
   What was read, watched, seen, listened to? And consider the ratio between the mediums.

1.

2.

3.

4.

WEEK FOUR

5.

6.

7.

8.

CULTURE

THERE ARE REALLY two kinds of life. There is, as Viri says, the one people believe you are living, and there is the other. It is this other which causes the trouble, this other is to see.

—James Salter, *Light Years*

Define authentic desire.

WEEK ONE

TAKE A RECENT INVENTORY

1. A sight:
2. A smell:
3. A sound:
4. A touch:
5. A taste:

Crystal balls, 1908–1913.
The New York Public Library Digital Collections.

1.

2.

3.

4.

5.

Remember last night.

23 / 54

WEEK TWO

WHO IS THE YOU OF NOW?

A survey of present interests, curiosities, recurring themes . . . a space for a short list of some front-burner items:

1. Thinking small
2. Rosehip seed oil
3. Hotel staycations
4. Marzipan
5. The Pixies

1.

2.

3.

4.

5.

Moon model prepared by Johann Friedrich Julius Schmidt. Constructed of 116 sections of plaster on a framework of wood and metal. Germany, 1898.
The Field Museum Library.

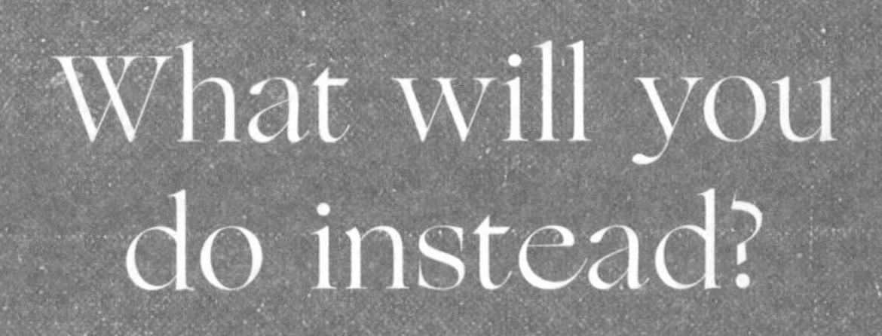

24 / 54

WEEK THREE

A SPACE FOR SCRAPS

A space to preserve. Tape or insert something (or things) you've acquired this month, a tactile reminder of the present.

The Past Month

1. THEMES
   Were there any themes last month?

2. UTILITY
   Describe a moment last month when you felt useful.

3. MINOR SECRETS
   Describe something you did in private. Perhaps it's not really a secret but something that never occurred to you to share. . . .

4. ADJUST
   What are you amid that is almost (but not quite) right? A draft, a relationship, an injury . . . what needs refinement and attention?

5. WEIRD
   When did you marvel at the peculiar?

6. OBSCURE LOVE
   What did you love (or discover) on your own? Maybe you remembered something you once loved.

7. NATURE
   An encounter with and/or in the outdoors. Connection with the natural world.

8. CULTURE LIST
   What was read, watched, seen, listened to? And consider the ratio between the mediums.

1.

2.

3.

4.

WEEK FOUR

5.

6.

7.

8.

OBSCURE LOVE

I . . . REALLY LOVE corn chips.

Eventually, feelings fade.

WEEK ONE

TAKE A RECENT INVENTORY

1. A sight:
2. A smell:
3. A sound:
4. A touch:
5. A taste:

Porcelain Moon Jar from the Joseon Dynasty, mid-18th century. Named for shape and milky glazing, and traditionally made from two hemispherical halves joined together in the center. *Metropolitan Museum of Art.*

1.

2.

3.

4.

5.

Call a friend.

WEEK TWO

WHO IS THE YOU OF NOW?

A survey of present interests, curiosities, recurring themes . . . a space for a short list of some front-burner items:

1. The combination of mango and basil
2. Holding hands
3. Palo Santo wood
4. The Bridge of Immortals
5. Peggy Guggenheim

1.

2.

3.

4.

5.

Faience Egyptian Ring,
ca. 1550–1295 B.C.
Metropolitan Museum of Art.

Maybe transition requires a change of location.

27 / 54

WEEK THREE

A SPACE FOR SCRAPS

A space to preserve. Tape or insert something (or things) you've acquired this month, a tactile reminder of the present.

The Past Month

1. PALETTE
   Give the last month a color (or a combination of possible swatches).

2. SCENT
   Give the last month a smell.

3. TASTE
   Give the last month a taste.

4. ALONE TIME
   What role did alone time play? Recount an experience.

5. AHA
   What was a breakthrough? A step forward in a learning process, or perhaps a way of seeing something in a new way?

6. GRATITUDE
   What were you particularly thankful for?

7. VULNERABILITY
   When were you exposed, raw, self-conscious?

8. CULTURE LIST
   What was read, watched, seen, listened to? And consider the ratio between the mediums.

1.

2.

3.

4.

WEEK FOUR

5.

6.

7.

8.

SCENT

A NEIGHBORHOOD RESTAURANT I like has its own custom candle that burns in the bathroom. I've always liked the smell—it's a little unusually savory, almost like pastis. On a whim, I packed the candle in my bag when I went on a weeklong work trip last month, lighting it in my hotel room when I would come back in the evenings. A small thing, but it reminded me of home.

It's okay that maybe you don't want what someone else wants. Keep something for yourself.

WEEK ONE

TAKE A RECENT INVENTORY

1. A sight:
2. A smell:
3. A sound:
4. A touch:
5. A taste:

Decorative plate by William Chaffers, London, 1872.
The J. Paul Getty Museum.

1.

2.

3.

4.

5.

Keep it casual.

29 / 54

WEEK TWO

WHO IS THE YOU OF NOW?

A survey of present interests, curiosities, recurring themes . . . a space for a short list of some front-burner items:

1. Simple gestures
2. Mallorca
3. Taffy
4. Concrete floors
5. Donald Judd's furniture

1.

2.

3.

4.

5.

Model of planets in orbit around Trylon and Perisphere, 1935–1945.
The New York Public Library Digital Collections.

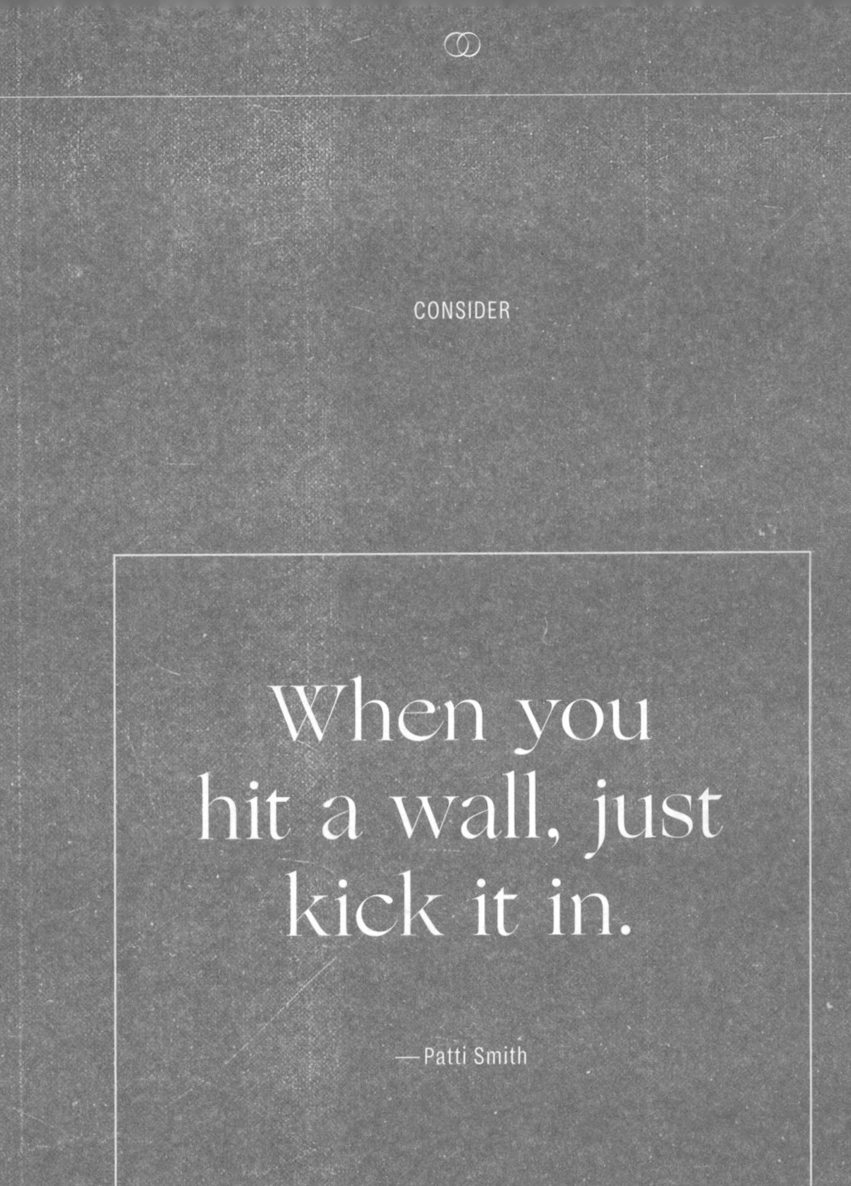

When you hit a wall, just kick it in.

—Patti Smith

30 / 54

WEEK THREE

A SPACE FOR SCRAPS

A space to preserve. Tape or insert something (or things) you've acquired this month, a tactile reminder of the present.

The Past Month

1. ENRICH
   What filled you up? A person? A meaningful meal? A place?

2. STIRRED
   What moved you? What stirred emotion (perhaps deeper than you expected)?

3. DISCARD
   What is something to rid from your life (or that you got rid of last month)?

4. ANNOYED
   Did you snap? What frustrated you or wore on your patience?

5. FUNNY
   What was a funny thing that happened? When did you laugh most deeply? When did you smile like a fool? Recount the experience.

6. CLARITY
   What came into focus for you?

7. FLOW
   Was there a moment when you lost yourself and/or lost time in pursuit of an activity or idea?

8. CULTURE LIST
   What was read, watched, seen, listened to? And consider the ratio between the mediums.

1.

2.

3.

4.

WEEK FOUR

5.

6.

7.

8.

CLARITY

THE LAST COUPLE months have felt foggy, unfocused. I have been working too much. I've amassed a small collection of tinctures, sprays, oils, balms . . . all infused with healing, consciousness-heightening ingredients. Apparently, I can't resist a good tincture. I think I actually just need to take some time off.

CONSIDER

Sometimes you get simple things wrong.

31 / 54

WEEK ONE

TAKE A RECENT INVENTORY

1. A sight:
2. A smell:
3. A sound:
4. A touch:
5. A taste:

Astrological Chart, 1877.
The New York Public Library Digital Collections.

1.

2.

3.

4.

5.

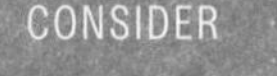

CONSIDER

What are your options?

32 / 54

WEEK TWO

WHO IS THE YOU OF NOW?

A survey of present interests, curiosities, recurring themes . . . a space for a short list of some front-burner items:

1. Hot springs
2. White linen
3. A Kit Kat bar
4. Found flowers
5. Manifestos

1.

2.

3.

4.

5.

Spouted jug with raised concentric circles from Northwestern Anatolia, ca. 2700–2400 B.C. *Metropolitan Museum of Art.*

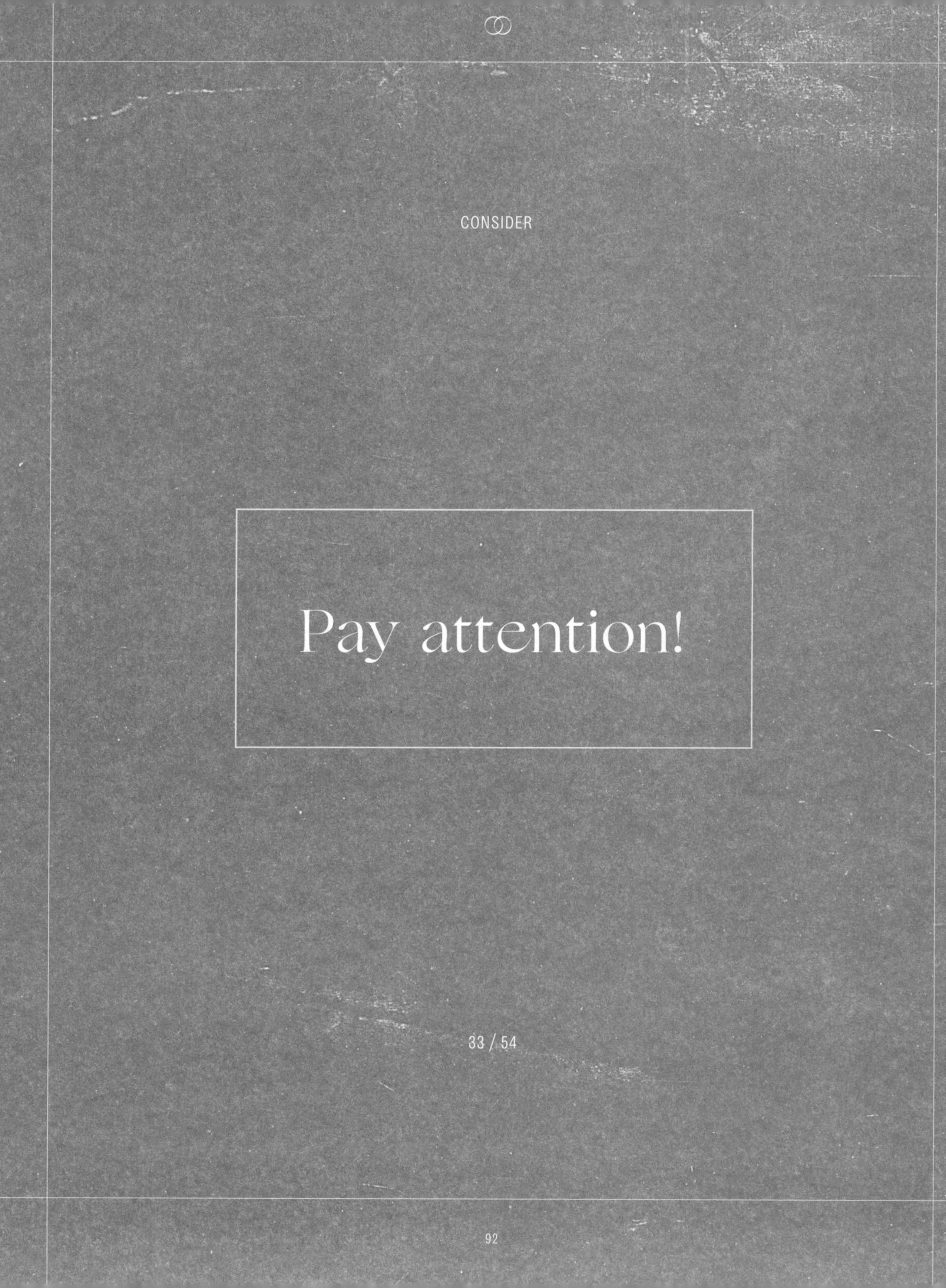

33 / 54

WEEK THREE

A SPACE FOR SCRAPS

A space to preserve. Tape or insert something (or things) you've acquired this month, a tactile reminder of the present.

The Past Month

1. THROES
   What are you in the midst of? Maybe: work, solitude, connection, nausea, transition, small pleasures, helplessness . . . ? Pick a word. Describe it.

2. CLOTHING
   What did you wear? Are there particular garments that are having a moment for you?

3. CURIOUS
   What piqued intrigue?

4. CHANGE
   A change you enacted with intention, or one that came from the outside.

5. LONELY
   What role did loneliness play?

6. ART EXPERIENCE
   An encounter with art (in any form).

7. MOST USED
   Take an inventory of what physical things defined the last month: a bobby pin to hold back your new bangs, a paper bag of just-ripe figs, a new pen, a balm for sunburned lips . . .

8. CULTURE LIST
   What was read, watched, seen, listened to? And consider the ratio between the mediums.

1.

2.

3.

4.

WEEK FOUR

5.

6.

7.

8.

CLOTHING

A FRIEND GAVE me this straw sun hat. Historically I've not been one for hats (I actually think this might somehow be the first one I've ever owned), and the newness is still novel. It feels like a disguise, another identity to pop on every once in a while.

Sometimes you don't get to choose.

WEEK ONE

TAKE A RECENT INVENTORY

1. A sight:
2. A smell:
3. A sound:
4. A touch:
5. A taste:

A map of the Earth's Western Planisphere, ca. 1757.
The New York Public Library Digital Archives.

1.

2.

3.

4.

5.

Just sit there.

35 / 54

WEEK TWO

WHO IS THE YOU OF NOW?

A survey of present interests, curiosities, recurring themes . . . a space for a short list of some front-burner items:

1. Earthship houses
2. Pina Bausch
3. Exfoliation
4. A Peter Pan collar
5. Tagliatelle

1.

2.

3.

4.

5.

Practice of Balloon Fleet, 1896.
The New York Public Library Digital Collections.

Allow a good idea to outrun you.

WEEK THREE

A SPACE FOR SCRAPS

A space to preserve. Tape or insert something (or things) you've acquired this month, a tactile reminder of the present.

The Past Month

1. LEARNING CURVE
   What needs some practice?

2. JOY
   A time of celebration or deep happiness, or that made you feel emotionally full.

3. ABSENCE
   What was missing last month?

4. FLASHY
   What did you do to excess? (And was it fun?)

5. SEASON AND STORY
   What is it presently a season for, and how was a ritual or experience informed by the time of year? A special outdoor dinner in the spring. A fresh peach in the summer. A snowed-in and housebound Saturday.

6. CLARITY
   What came into focus for you?

7. NOSTALGIA
   A moment that transported you to your own past.

8. CULTURE LIST
   What was read, watched, seen, listened to? And consider the ratio between the mediums.

1.

2.

3.

4.

WEEK FOUR

5.

6.

7.

8.

SEASON AND STORY

IT'S HIGH SUMMER. Tomato season. My dad has been periodically dropping off cardboard boxes filled with the latest harvest from his backyard garden, and I can't eat them fast enough. Nearly every day I've had one in the afternoon, sliced thinly and topped with za'atar, a savory Middle Eastern spice mix of sesame, thyme, and sumac. It's a proper dose of depth amid the summer's lightness.

Avoid what reduces you.

Heath sculp

WEEK ONE

TAKE A RECENT INVENTORY

1. A sight:
2. A smell:
3. A sound:
4. A touch:
5. A taste:

"Nautilus Pompilius or Pearly Nautilus; Longitudinal Section of the Shell to shew [sic] the internal structure," 1809.
The New York Public Library Digital Collections.

1.

2.

3.

4.

5.

CONSIDER
Let light in.
38 / 54

WEEK TWO

WHO IS THE YOU OF NOW?

A survey of present interests, curiosities, recurring themes . . . a space for a short list of some front-burner items:

1. French chiffon
2. David Lynch
3. Baroque pearls
4. Gin rummy
5. Vibrations

1.

2.

3.

4.

5.

Wedding dinner menu, Cincinnati, Ohio, 1892.
The New York Public Library Digital Collections.

CONSIDER

If you feel ambivalent, it is best to just be patient. Wait.

39 / 54

WEEK THREE

A SPACE FOR SCRAPS

A space to preserve. Tape or insert something (or things) you've acquired this month, a tactile reminder of the present.

The Past Month

1. OBJECT
   What was an unexpected object that entered your life? A bouquet of helium-filled balloons. A worm. A tumbleweed.

2. RESIST
   What did you resist or push back against?

3. STIRRED
   What moved you? What stirred emotion (perhaps deeper than you expected)?

4. RETURN
   What did you revisit last month? A song you played over and over. A film you rewatched for the first time in years. A breakfast rut.

5. FISH OUT OF WATER
   When were you out of place or out of your element?

6. NATURE
   An encounter with and/or in the outdoors. Connection with the natural world.

7. THEMES
   Were there any themes last month?

8. CULTURE LIST
   What was read, watched, seen, listened to? And consider the ratio between the mediums.

1.

2.

3.

4.

WEEK FOUR

5.

6.

7.

8.

NATURE

ON THE HOTTEST day of the year (so far) I went on an afternoon walk. Sweaty and tired, at some point I tripped and twisted my ankle. Sitting on the side of the curb, I strategized how I was going to get back since I was still a couple of miles away. Suddenly there was a gust of wind and raindrops started to fall, one of those random summer storms where a single dark cloud seems to drop rain directly over you, even when the rest of the sky is blue. I started laughing, imagining the moment as a real-life Charlie Brown comic. I picked up a leafy branch and used it as a makeshift umbrella, then got up and limped home.

Is it in your mind, or is it real?

WEEK ONE

TAKE A RECENT INVENTORY

1. A sight:
2. A smell:
3. A sound:
4. A touch:
5. A taste:

Partial eclipse of the moon.
Observed October 24, 1874.
The New York Public Library Digital Collections.

1.

2.

3.

4.

5.

Rediscover your boundaries.

41 / 54

WEEK TWO

WHO IS THE YOU OF NOW?

A survey of present interests, curiosities, recurring themes . . . a space for a short list of some front-burner items:

1. Scales of justice
2. A Japanese comb
3. Savory fruit salad
4. Willow baskets
5. Rice pudding

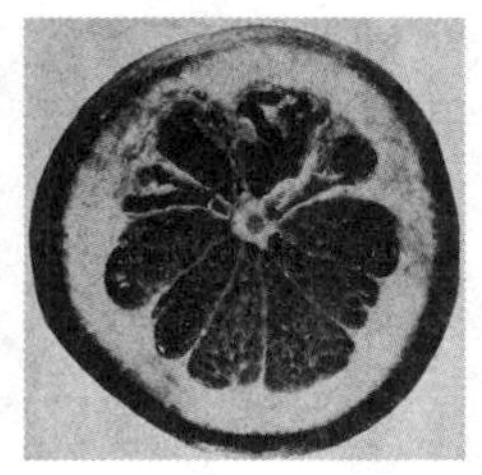

1.

2.

3.

4.

5.

"Bulletin of the U.S. Department of Agriculture," 1913.
Smithsonian Libraries.

Not all dreams need to be realized.

42 / 54

WEEK THREE

A SPACE FOR SCRAPS

A space to preserve. Tape or insert something (or things) you've acquired this month, a tactile reminder of the present.

The Past Month

1. ADJUST
   What are you amid that is almost (but not quite) right? A draft, a relationship, an injury . . . what needs refinement and attention?

2. NATURE
   An encounter with and/or in the outdoors. Connection with the natural world.

3. CHANGE
   A change you enacted with intention, or one that came from the outside.

4. LOSS
   What was lost? What are you mourning? Maybe it is minor.

5. CONVERSATION
   A meaningful dialogue.

6. TAKEN FOR A RIDE
   In retrospect, was there an interaction, experience, or situation where you were taken advantage of (intentional or not)?

7. MYSTERY
   What happened that you can't explain?

8. CULTURE LIST
   What was read, watched, seen, listened to? And consider the ratio between the mediums.

1.

2.

3.

4.

WEEK FOUR

5.

6.

7.

8.

CHANGE

I GOT A HAIRCUT. I've never styled my hair or even blow-dried it regularly, but after realizing that my new chop caused the bottoms of my hair to flip outward in a way that deeply annoyed me . . . I bought my very first hair straightener. I like the haircut much better now.

It is okay to love something obvious.

WEEK ONE

TAKE A RECENT INVENTORY

1. A sight:
2. A smell:
3. A sound:
4. A touch:
5. A taste:

Miroir étrusque, 1880–1883.
The New York Public Library Digital Collections.

1.

2.

3.

4.

5.

Expand your range of what is a possibility.

WEEK TWO

WHO IS THE YOU OF NOW?

A survey of present interests, curiosities, recurring themes . . . a space for a short list of some front-burner items:

1. A sun hat
2. Mallard
3. Grilled cheese
4. Susan Sontag
5. Déjà vu

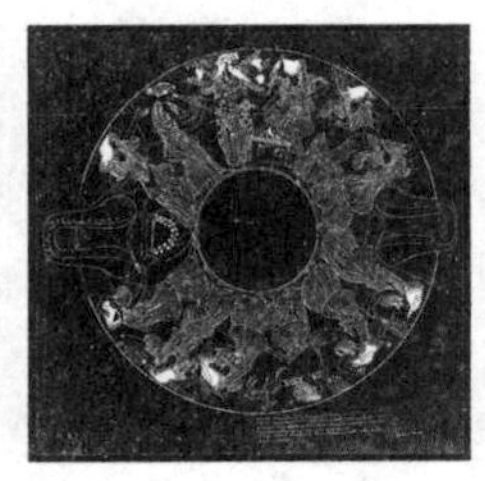

1.

2.

3.

4.

5.

"Maenads with thyrsi dance in a ring around the ancient wooden statue of Dionysus," 1894. *The New York Public Library Digital Collections.*

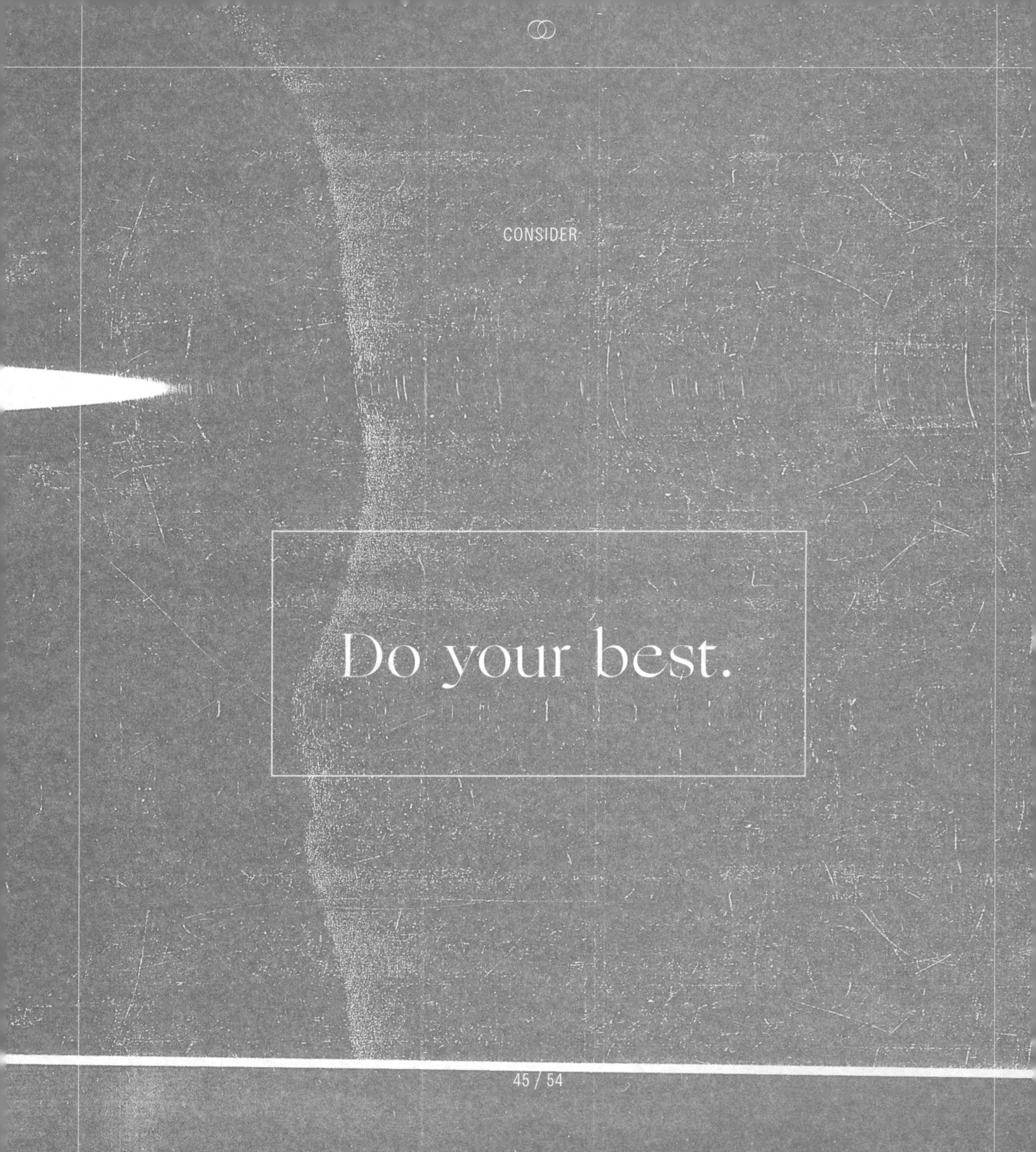
CONSIDER
Do your best.
45 / 54

WEEK THREE

A SPACE FOR SCRAPS

A space to preserve. Tape or insert something (or things) you've acquired this month, a tactile reminder of the present.

The Past Month

1. WEIRD
   When did you marvel at the peculiar?

2. ENRICH
   What filled you up? A person? A meaningful meal? A place?

3. GIFT
   Something you gave or received.

4. REARRANGE
   Did you put something in a new order? Maybe something suits you better at a different time of day; maybe you organized a desk drawer.

5. PHYSICAL SELF
   Take a scan. What happened to your body? How do you feel?

6. PARE BACK
   Did you simplify something?

7. PROPORTION
   A specific moment that reminded you of the scale of the universe, that you are part of a greater whole. An outdoor shower in the mountains. A particular drive at night. Reading a specific quotation at the right time.

8. CULTURE LIST
   What was read, watched, seen, listened to? And consider the ratio between the mediums.

1.

2.

3.

4.

WEEK FOUR

5.

6.

7.

8.

REARRANGE

I TOOK INVENTORY of an overflowing pantry and finally put all of my dried grains, beans, and nuts into the glass jars I bought nearly a year ago. Small acts of order felt empowering.

CONSIDER

Abandon one story you are telling and develop a new one.

46 / 54

WEEK ONE

TAKE A RECENT INVENTORY

1. A sight:
2. A smell:
3. A sound:
4. A touch:
5. A taste:

Egyptian Papyrus Lid,
ca. 1336–1327 B.C.
Metropolitan Museum of Art.

1.

2.

3.

4.

5.

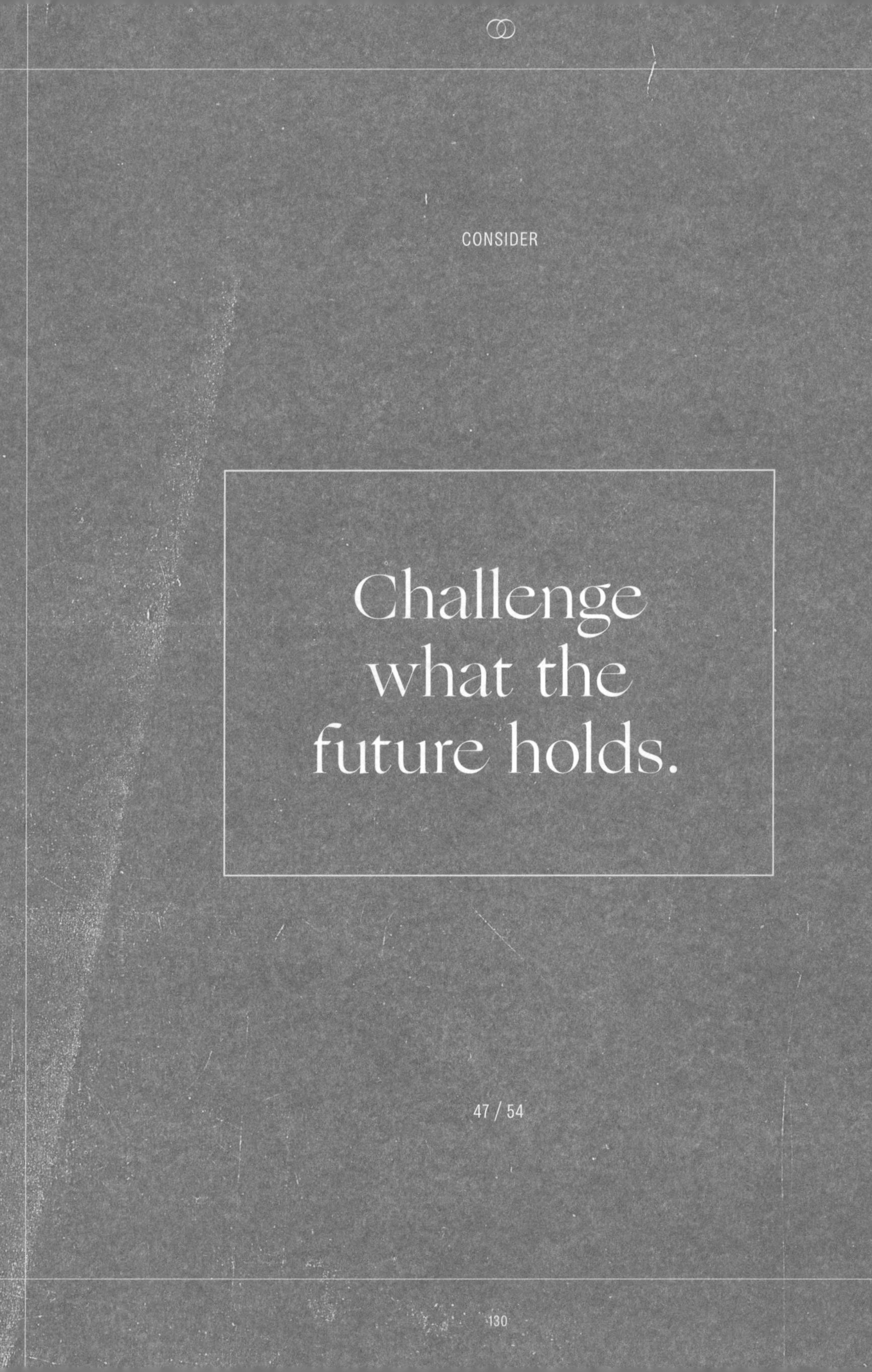

Challenge what the future holds.

47 / 54

WEEK TWO

WHO IS THE YOU OF NOW?

A survey of present interests, curiosities, recurring themes . . . a space for a short list of some front-burner items:

1. Opalescence
2. Mechanical pencils
3. Fig season
4. Gifts for no reason
5. Meringue

1.

2.

3.

4.

5.

19th century metal eyeglasses.
Metropolitan Museum of Art.

Maybe the thing you are most interested in has always been there.

WEEK THREE

A SPACE FOR SCRAPS

A space to preserve. Tape or insert something (or things) you've acquired this month, a tactile reminder of the present.

The Past Month

1. SIMPLE NOURISHMENT
   What simply satisfied you? A ten-minute phone call with a friend. A perfectly timed bath. A sandwich.

2. WONDER
   What moments crossed the threshold in making you wonder? When did you feel amazement at the world around you or experience a marked break from your routine?

3. UTILITY
   When were you useful?

4. NOSTALGIA
   What was felt more deeply because it took you back to your past?

5. CLARITY
   What came into focus for you?

6. UNEXPECTED PAIRINGS
   What nonobvious combinations had a role: two random songs, played sequentially and on repeat; a peach and a tomato, together in a salad.

7. BRAVERY
   What took courage?

8. CULTURE LIST
   What was read, watched, seen, listened to? And consider the ratio between the mediums.

1.

2.

3.

4.

WEEK FOUR

5.

6.

7.

8.

UTILITY

WHILE WE WERE scrambling to light the grill on a camping trip, I figured out how to light a match on my pant zipper. Useful, and it's a good party trick!

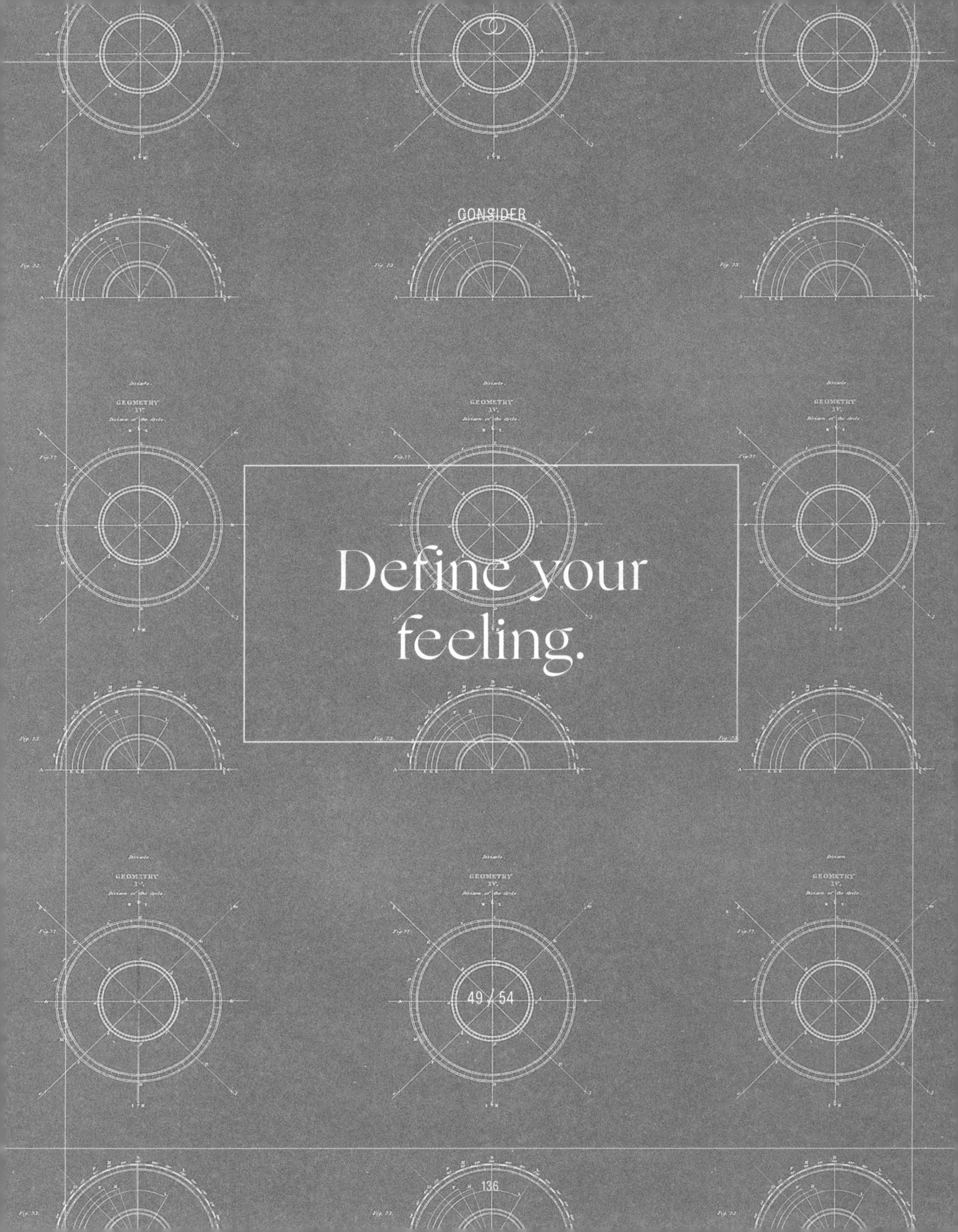

Define your feeling.

49 / 54

WEEK ONE

TAKE A RECENT INVENTORY

1. A sight:
2. A smell:
3. A sound:
4. A touch:
5. A taste:

"Geometry IV. Division of the Circle," 1826.
The New York Public Library Digital Collections.

1.

2.

3.

4.

5.

CONSIDER

Planning ahead sometimes works.

50 / 54

WEEK TWO

WHO IS THE YOU OF NOW?

A survey of present interests, curiosities, recurring themes . . . a space for a short list of some front-burner items:

1. Pisa
2. Grapefruit juice
3. Tulle
4. Child prodigies
5. Seaweed

1.

2.

3.

4.

5.

Sketch of observatory at the New York World's Fair, 1935–1945.
The New York Public Library Digital Collections.

Feel your fingers.

WEEK THREE

A SPACE FOR SCRAPS

A space to preserve. Tape or insert something (or things) you've acquired this month, a tactile reminder of the present.

The Past Month

1. NARRATIVE
   You run into a friend you haven't seen in a while—"What's been going on with you?" What's the story you're currently telling about yourself, to yourself (and to others)?

2. OBJECT
   An object (New? Old?) that played a role in your life last month.

3. CHALLENGE
   What was a question you found yourself asking?

4. NATURE
   An encounter with and/or in the outdoors. Connection with the natural world.

5. AHA
   What was a breakthrough? A step forward in a learning process, or perhaps a way of seeing something in a new way?

6. MESS
   What was messy?

7. CONFIDENCE
   What was a moment where you felt at ease with yourself?

8. CULTURE LIST
   What was read, watched, seen, listened to? And consider the ratio between the mediums.

1.

2.

3.

4.

WEEK FOUR

5.

6.

7.

8.

OBJECT

I WENT TO an estate sale in my neighborhood. I love the voyeurism that comes from wandering through a stranger's house, imagining the lives their objects lived and the order with which the original owner arranged their world. There is, I suppose, a sadness to it all, but in the moment it mostly fills me with curiosity for the stories that remain beyond us. Typically, I leave estate sales with books or ceramics, but this time I was drawn to a glass orb. It's startlingly heavy, taking two hands to properly handle it; now in my space, it casts prismatic rainbows around my office when the light is just right.

CONSIDER
Be beyond.
52 / 54

WEEK ONE

TAKE A RECENT INVENTORY

1. A sight:
2. A smell:
3. A sound:
4. A touch:
5. A taste:

Egyptian gold and carnelian necklace, ca. 1550–1295 B.C. *Metropolitan Museum of Art.*

1.

2.

3.

4.

5.

What are you resisting?

WEEK TWO

WHO IS THE YOU OF NOW?

A survey of present interests, curiosities, recurring themes . . . a space for a short list of some front-burner items:

1. Impulse buys
2. Joshua Tree
3. Almond oil
4. Cerulean
5. The law of attraction

1.

2.

3.

4.

5.

Shell Gorget—The Bird,
Mississippi, 1883.
The New York Public Library Digital Collections.

Just show up.

54 / 54

WEEK THREE

A SPACE FOR SCRAPS

A space to preserve. Tape or insert something (or things) you've acquired this month, a tactile reminder of the present.

The Past Month

1. STORY
   What was the best story of the last month? Maybe it was something that happened to you. Maybe it was something that was told to you or that you observed.

2. STOMACH SINKING
   When did your body flag an emotional reaction—a dip in the road, the realization of error, remembering that needles make you nauseous?

3. TIME OF DAY
   What times felt significant: Dawn or dusk? A regular hour that brings lull or excitement? A time you looked forward to or anticipated?

4. ENRICH
   What filled you up? A person? A meaningful meal? A place?

5. DISCARD
   What did you get rid of?

6. SURPRISE
   What was shocking (perhaps delightfully so)?

7. REARRANGE
   Did you put something in a new order? Maybe something suits you better at a different time of day; maybe you organized a desk drawer.

8. CULTURE LIST
   What was read, watched, seen, listened to? And consider the ratio between the mediums.

1.

2.

3.

4.

WEEK FOUR

5.

6.

7.

8.

TIME OF DAY

THIS MONTH, I was intentional about sitting down to have a proper dinner at home. Even if we were eating from takeout boxes, I would light this candle at the table. The act of striking the match, lighting the candle, and sitting down to mark the day's end brought a certain satisfaction.

Look back at the first few lists in this book. What has changed? Where do you lie between yearning for the future and feeling nostalgic for what's passed? Consider that, and move ahead.

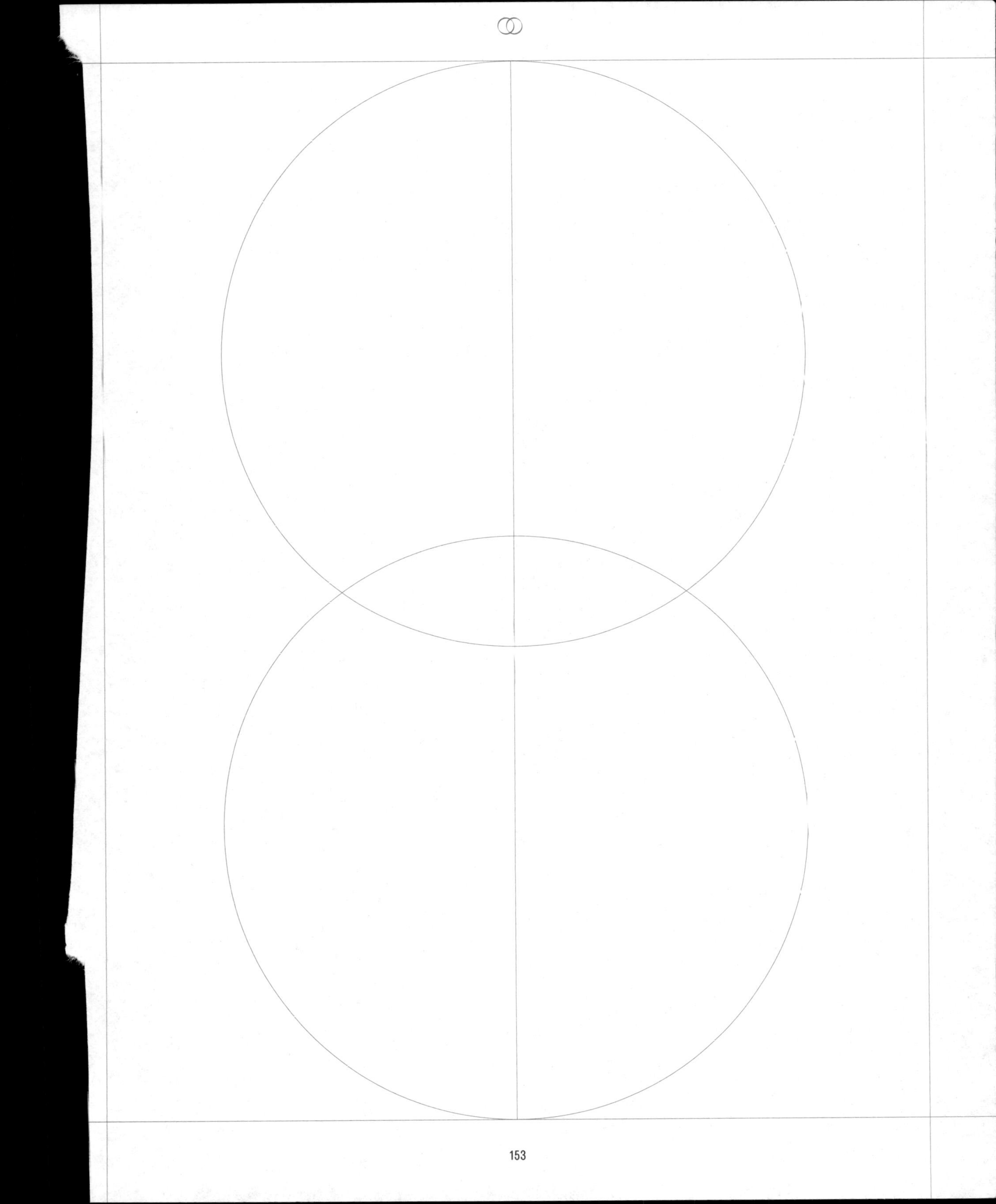

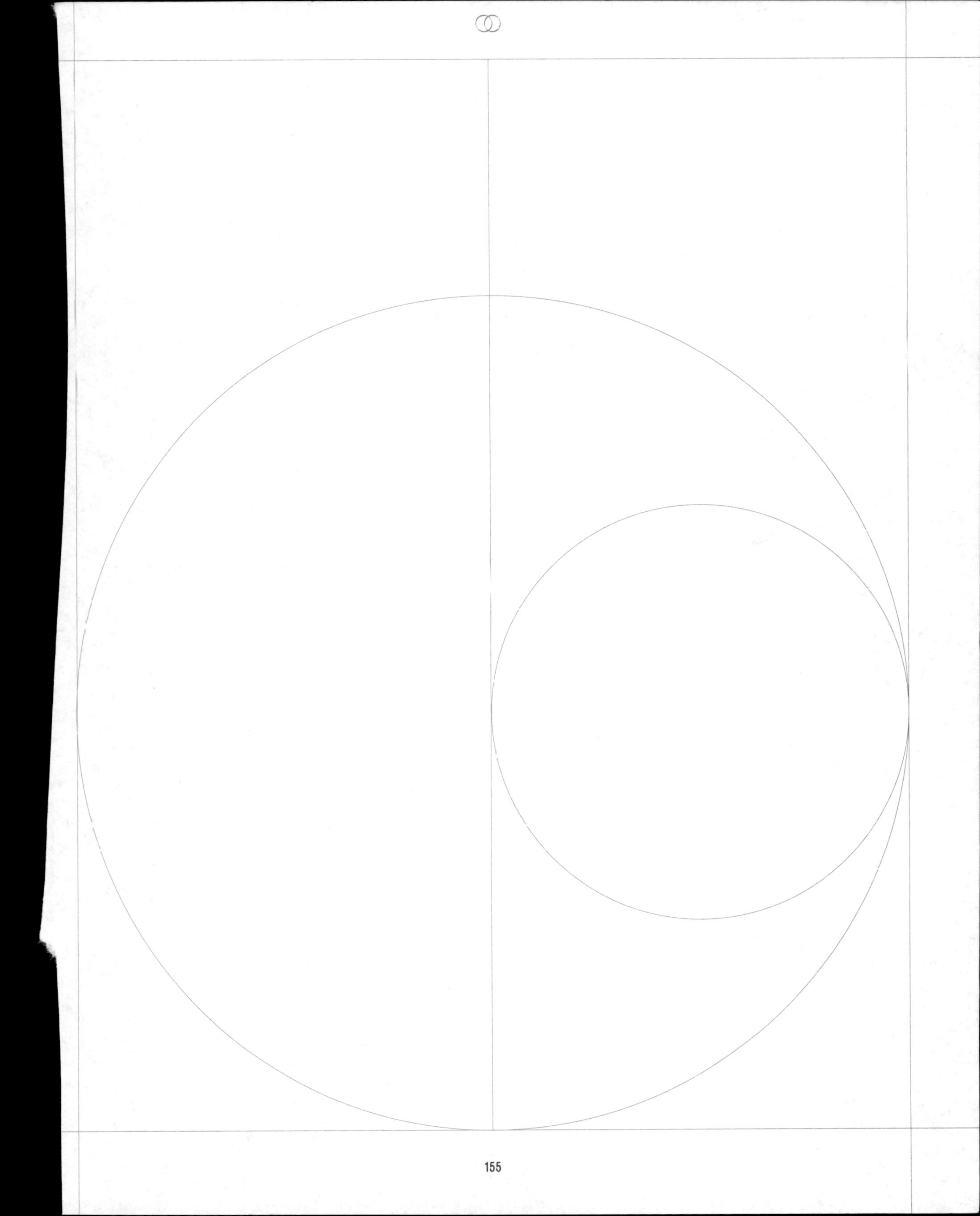

Published in the United States by Clarkson Potter/Publishers, an imprint of the Crown Publishing Group, a division of Penguin Random House LLC, New York.
crownpublishing.com
clarksonpotter.com

CLARKSON POTTER is a trademark and POTTER with colophon is a registered trademark of Penguin Random House LLC.

ISBN 978-1-9848-2272-7

Printed in China

Book and cover design by Natasha Mead
Still life object photography by Alyson Fox
Textures by texturefabrik.com

Photographs on page 5: (*left*) "*Dancing with Helen Moller*; her own statement of her philosophy and practice and teaching formed upon the classic Greek model, and adapted to meet the aesthetic and hygienic needs of today," p. 44. 1918. *University of California Libraries*; (*middle*) Pewter Sundial, 1762. *The New York Public Library Digital Collections*; (*right*) Terracotta flask with ring-shaped body, ca. 850–600 B.C. *Metropolitan Museum of Art.*

10 9 8 7 6 5 4 3 2

First Edition